The Short Vertical Antenna and Ground Radial

By Jerry Sevick, W2FMI

CQ Communications, Inc.

Library of Congress Control Number 2003101976
ISBN 0-943016-22-3

Editor: Edith Lennon, N2ZRW
Layout and Design: Elizabeth Ryan
Illustrations: Hal Keith

Published by CQ Communications, Inc.
25 Newbridge Road
Hicksville, New York 11801 USA

Printed in the United States of America.

Copyright © 2003 CQ Communications, Inc., Hicksville, New York

All rights reserved. No part of this book may be used or reproduced in any form, or stored in a database or retrieval system, without express written permission of the publisher except in the case of brief quotes included in articles by the author or in reviews of this book by others. Making copies of editorial or pictorial content in any manner except for an individual's own personal use is prohibited, and is in violation of United States copyright laws.

While every precaution has been taken in the preparation of this book, the author and publisher assume no responsibility for errors or omissions. The information contained in this book has been checked and is believed to be entirely reliable. However, no responsibility is assumed for inaccuracies. Neither the author nor the publisher shall be liable to the purchaser or any other person or entity with respect to liability, loss, or damage caused or alleged to have been caused directly or indirectly by this book or the contents contained herein.

This book is dedicated to my late wife, Connie.

Preface

A job transfer in 1969 from the Branch Lab at the Western Electric Plant in Allentown, Pennsylvania, to the Bell Labs Headquarters at Murray Hill, New Jersey, made a large impact on my technical future. At Bell Labs, I went from supervising a group in Transistor Development to being a Director of a center named Technical Relations. The center dealt with exchanges between our technical staff and the outside world.

My hobby, Amateur Radio, also led to experiences in different technologies. After my move to New Jersey, I did not erect my 40-foot tower holding a beam antenna. This was a good neighborly gesture as well as a good chance to investigate the short vertical antenna and ground systems. This resulted in my writing articles for various publications and, eventually, to my study of the transmission line transformer (TLT).

My first interest in the short vertical antenna began in 1951 when my thesis advisor at Harvard, Professor R.W.P. King, developed the full theory of the short antenna. In it he disclosed that a very short antenna had about the same power gain and radiation pattern as a full size half-wave antenna. The main difference was that the resistive component of the input impedance, the radiation resistance, was very small in comparison to that of a full 1/2-wavelength antenna and, depending upon length, could be a matter of a few ohms. In turn, the short antenna has a very high capacitive reactance, which has to be canceled by various loading techniques. A short vertical has an even lower radiation resistance and, depending upon height, can be a matter of only 1 or 2 ohms.

A literature search I conducted found very little information on short verticals and ground radials in the amateur and professional radio publications. In the amateur realm, three-foot ground rods and four 1/4-wavelength-long ground radials were casually mentioned. In the professional literature, I found only one publication. It was written by three scientists from RCA Corporation—Brown, Lewis, and Epstein—and was published in the 1937 *Proceedings of the IRE*. I, therefore, decided to investigate the number of radials really needed for short verticals.

My first attempt was a study of ground radials placed on the surface of the earth. This was unlike the RCA study, which involved burying the radials one foot below the surface. After this, I investigated the various loading techniques, including top hat, mid point, base, and helical loading. My final design was for a six-foot antenna for 40 meters, which included a high degree of top hat loading and gave the highest radiation resistance. The final structure was a beach umbrella antenna and is shown, with my late wife comfortably sitting under it, in Chapter 2. Needless to say, it gave an outstanding performance since there was essentially no loss in the 100 radials I used.

This book, which details my work in this area, is the result of a recent phone conversation with Dick Ross, publisher of CQ Communications. While in the process of developing a simplified and unified theory of the TLT for a Video/CD of my book *Transmission Line Transformers*, published by Noble Publishing, I asked Dick for permission to use several figures and photos from my previous work under his publication, *Building and Using Baluns and Ununs*. When he found out what the request was for, he immediately suggested that I update that book as well. Thus several changes took place. The title of that work was changed to *Understanding, Building, and Using Baluns and Ununs* and the Appendices on the "The Short Vertical Antenna and Ground Radial" were culled to create this book. Obviously, this information didn't fit in its original form in *Building and Using Baluns and Ununs* because it was not included in its title and, therefore, not available through the usual search engines.

In closing, I refer you to the concluding statements in Chapter 1. Through many business arrangements I got to know Dr. Brown quite well. On one occasion, while waiting for our flights at Newark Airport, I asked him how he and his associates arrived at 120 radials. I had just finished my study on ground radials, which showed that 100 were sufficient. His answer shows that there can be humor in the technical world.

And, finally, I'd like to credit some who have made it possible to more easily present my results on the short vertical antenna and the latest views on the broadband technology of the TLT. They are Dick Ross, publisher CQ Communications, Edith Lennon, Editor, and my associate at Basking Ridge, Kathy Collyer.

Jerry Sevick, W2FMI
Fellowship Village
Basking Ridge, New Jersey
January 29, 2003

Table of Contents

Preface v

Chapter 1
    Ground-Radial System for Verticals 1

Chapter 2
    Short Ground-Mounted Verticals 9

Chapter 3
    Short Ground-Radial Systems for
    Short Verticals 23

Chapter 4
    The Loading Coil 37

Index I-1

Chapter 1

Ground-Radial System for Verticals

Many years ago, when I moved to a new neighborhood, I found myself looking at my antenna needs in a new light. I wanted to minimize or, at best, avoid any obstacles to good neighborly relations by not immediately reinstalling my triband Yagi beam on its 40-foot tower. A ground-mounted vertical beam antenna (i.e., an array of vertical anten-

Photo 1-A. *The experimental setup I eventually used in my vertical (and other ground-fed) antenna studies. It has a large aluminum plate (10 to 15 inches in diameter), a resistive bridge with external meter, and a variable signal source. In this case, the transceiver in the shack is the signal source.*

nas coupled together with the appropriate feed system) seemed the logical choice for my new installation because it would:

1) Exhibit a low profile.
2) Not require extra help in installation.
3) Offer a low angle of directed radiation.
4) Not require a large outlay of money.
5) Give me some experience with vertical antennas.

Consequently, I constructed a square array of four vertical antennas with four 1/4-wave radials (on the surface of the earth) under each vertical. When my tests began, it was immediately apparent that the simple procedures I was using were inadequate to cope with such a complex system. I had to start anew to develop a test procedure and build suitable test equipment. Because I didn't want to dig up my backyard for buried radials, and no practical information was available for radials on the earth's surface, the logical step was to backtrack to a single vertical antenna and use it as a model upon which to develop some basic standards. I found that a simple impedance bridge, a field-strength meter, and a test oscillator provided me with all of the data about verticals I needed. The results presented in this chapter are largely taken from my July 1971 *QST* article "The Ground-Image Vertical Antenna."

Ground versus Elevated Radials

A ground-image antenna differs considerably from a ground-plane antenna that relies merely on a few 1/4-wave radials above the ground. The ground-image antenna results when an adequate number of radials are used, so the image of only the vertical section is sufficient to describe it. Practical information on ground-image systems has appeared in the literature for many years.[1,2] The results have shown that some 100 1/2-wave radials buried just below the surface of the earth provide a good ground system. However, at higher frequencies, the dielectric effect of the earth brings about severe discrimination of radiation or reception at very low angles.[3,4]

The ground-plane vertical, which uses 1/4-wave radials (or odd-multiples thereof), can shield the vertical portion from the lossy earth depending upon the number of radials and the height above ground. It is apparent that the higher the antenna, the fewer the number of radials needed. The minimum number is three radials spaced 120 degrees apart. The disadvantage of this type of antenna at low frequencies (1.7 to 7.3 MHz) simply lies in the difficulty of constructing a system that has a sufficient number of radials and enough height to be an efficient radiator. Furthermore, if the installation is in your backyard, you might have a problem cutting the lawn!

For these reasons, I chose to investigate the effect of radials *just on* the surface of the earth and at frequencies greater than 3 MHz. The results of my findings were most gratifying. It became apparent that many "rules of thumb" which have been taken for granted over these many years were more myth than truth.

Experimental Results

The experimental setup I eventually used in my vertical (and other ground-fed) antenna investigations is shown in **Photo 1-A**. It has a large aluminum plate (10 to 15 inches in diameter) to which the radials are connected, a resistive bridge with an external meter, and a variable signal source. In the example shown, the signal source was the transceiver in my shack. Reading the classic paper by Brown, Lewis, and Epstein,[1] it appeared that a good starting point was to use radials 0.4 wavelengths long on the 20-meter band. Each radial, in turn, was terminated by large nails (10 to 12 inches long). These nails not only secured the

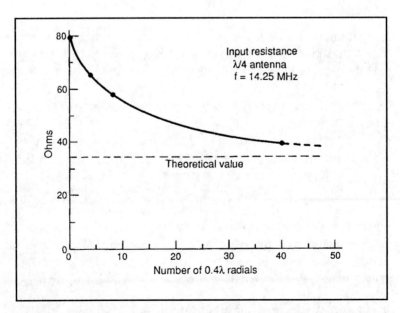

Figure 1-1. *The resonant input impedance of a 20-meter 1/4-wave vertical as a function of the number of 0.4-wavelength-long radials just on the surface of the earth. The 4 and 8 radials consisted of bundles of five No. 18 copper wires. The 40-radial point was obtained by fanning out the 8 radials.*

radial ends, they also provided an additional electrical contact with the earth.

The parameter I chose to measure was the resonant input impedance of a 1/4-wave vertical antenna. It is known to be:

$$R_{in} = R_{rad} + R_{loss} \qquad \text{(Eq 1-1)}$$

where:

R_{in} = the resonant input impedance
R_{rad} = the radiation resistance
R_{loss} = the loss in the ground system

In order to obtain some idea of the number of radials required for an adequate ground system, I used eight bundles of wire, each 25 feet long, and each made up of five No. 18 copper wires. I was then able to use each bundle as a radial, measure the input impedance, and then finally separate the wires in the bundles to arrive at 40 radials.

Figure 1-1 shows the resonant input impedance as a function of the number of radials. The results are surprisingly close to those of Brown, Lewis, and Epstein,[1] which were obtained with buried radials in the AM broadcast band. My technique points out some important features and refutes some of the old myths of ground systems:

1) Radials on the surface of the ground are as effective as buried radials. In fact, the radials should not be buried too deeply because the electric field only penetrates the earth a few feet at HF.

2) Four radials (even 0.4 wavelengths long) make a *very* poor ground system.

3) Forty radials were required at my location in order to provide an adequate ground system. Later work showed that 0.2 wavelength radials provided the same result (see **Chapter 2**).

4) Because each radial carries only 1/nth of the current, where n is the number of radials, and practically any metallic conductor is much better than the lossy earth, the type of conductor (copper, aluminum, or iron) and size is not very important. The main criterion is how long the radials will survive in the environment. This suggests that insulated wires could be the best way to go.

Another myth, which is frequently heard "on-the-air," is the idea that the minimum VSWR always occurs at the resonant frequency of an antenna. In fact, just the opposite is true. **Figure 1-1** shows that the radiation resistance of my vertical, which has a

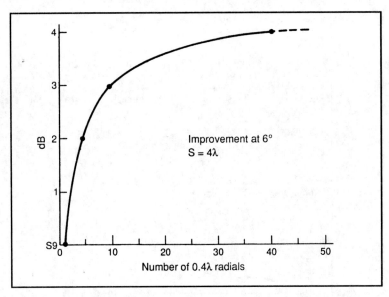

Figure 1-2. *The improvement in low-angle radiation of a 1/4-wave vertical on 20 meters as a function of the number of added 0.4-wavelength radials. A test oscillator was mounted on a wooden tower 4 wavelengths away at an elevation angle 6 degrees from the base of the vertical.*

height-to-radius ratio of 300,[5] is 35 ohms. Therefore, over a perfect ground system, the VSWR would be 1.4:1 at resonance. However, one finds that a minimum VSWR of 1.3:1 occurs a little above the resonant frequency. Even though the input impedance takes on an inductive component above resonance, the real part (the radiation resistance) also increases—resulting in the lower VSWR. Another misconception is that a 1:1 VSWR with a ground-mounted 1/4-wave vertical is an ideal condition. Again, not true. It shows that the ground-system loss is on the order of 15 ohms. With shortened verticals, the loss could even be greater.

Another aspect of this study was my investigation of the effect on the low-angle radiation as a function of the number and length of the radials. **Figure 1-2** shows the improvement in the low-angle radiation as a function of the number of 0.4-wavelength radials. I obtained these data by placing a test oscillator on a 26-foot wooden tower 4 wavelengths away from the 20-meter vertical.

The effect of using longer radials in a particular direction is shown in **Figure 1-3**. Radials of No. 18 wire, 3/2 wavelengths long, were placed between the existing 40 radials. The spacing between the longer wires was 5 degrees. The considerable

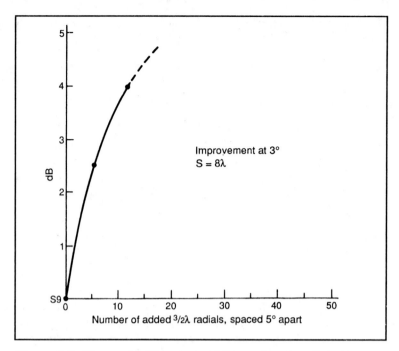

Figure 1-3. *The result of interlacing 3/2-wavelength radials in a particular direction. Data were obtained by using a test oscillator on a wooden tower 8 wavelengths away, at an elevation of 3 degrees from the base of the vertical. The results indicate the advantage of longer radials and the possibility of the directional properties of a nonsymmetrical radial system.*

improvement indicates the need for longer radials and the directional properties a ground plane could provide.

I also built and tested a 5/8-wavelength vertical. It consisted of a 40-foot telescoping aluminum pole and a loading coil of 8 turns of No. 12 wire with a diameter of 2.5 inches. A field strength meter was mounted at different heights on the 26-foot tower. The comparison in low-angle radiation at a distance of 7.5 wavelengths is presented in **Table 1-1**. The ground plane for these data consisted of the forty 0.4-wavelength radials plus the eleven 3/2-wavelength ones. Measurements were taken in the direction of the added longer radials. The results show the improvement in low-angle radiation offered by the 5/8-wavelength vertical. The resonant input impedance of this much taller vertical was found to be 76 ohms.

On-The-Air Checks

Taking these experiments and measurements into account, I settled on a 1/4-wavelength vertical with forty 0.4-wavelength (unburied) radials. I then proceeded to make on-the-air tests to compare the effectiveness with an inverted-V antenna having its apex at 0.4 wavelength, and with a 5/8-wavelength vertical over the same ground plane. Surprisingly, the 1/4-wavelength vertical seemed to perform just as well as the much taller 5/8-wavelength vertical. This could result from the fact that most signals arrive after several hops, and the optimum lobe angle is probably as high as 15 to 20 degrees.[6] At that angle, the 1/4-wavelength vertical actually enjoys an advantage. With practically all DX contacts, the verticals had a 6- to 8-dB improvement over the inverted-V! The only exceptions were at intermediate distances and for local contacts. At about 500 to 600 miles, the inverted-V with its higher angle of radiation gave better results. Locally, the verticals gave far superior performances. Improvements of 10 to 15 dB were recorded. I also tested a triband trap-vertical antenna on 20 meters and found it to be practically the same in impedance and performance as the 1/4-wavelength vertical. This antenna had an overall height of only 12.5 feet!

Closing Comments

The verticals in this study were matched to 50-ohm cable using L-C networks. At this time, I was only able to construct 4:1 Ununs matching 50-ohm cable to unbalanced loads of 12.5 ohms. However, simple fractional-ratio Ununs are now available to match 50-ohm cable to the impedances of the verticals in this study: namely, 35 and 76 ohms.

Finally, it should be mentioned that the world standard for the number of radials to be used with verticals in the AM broadcast

θ	$E(\lambda/4)$	$E(5/8\lambda)$	Gain of 5/8λ antenna
0.1 degree	0	0.58	∞
0.4 degree	0	0.62	∞
0.75 degree	0.62	0.80	3.4 dB
1.1 degrees	0.69	1.0	3.2 dB
1.5 degrees	0.62	0.92	3.4 dB
2.25 degrees	0.48	0.80	4.3 dB
3 degrees	0.41	0.69	4.5 dB

Table 1-1. *Comparison of responses of 1/4-wavelength and 5/8-wavelength vertical antennas at low radiation angles. Data were taken by a field-strength meter mounted on a wooden tower at a distance of 7.5 wavelengths and at 14.250 MHz. Field strength, E, is normalized to the maximum value obtained with the 5/8-wavelength case.*

band is 120. This number was based on the classic paper published in 1937 by Brown, Lewis, and Epstein.[1] During the course of a business meeting with Dr. Brown, I asked him how he and his colleagues arrived at the 120 radial figure—because I was quite sure 100 would work as well. His answer was interesting.

He said that he and the others had been thinking in terms of 100 radials, but the farmer who plowed in 100 radials had wire left over because copper is soft and stretches easily. When he asked what to do with the extra wire, the farmer was told to plow it in. The result was a world standard of 120 radials.

References

1. Brown, Lewis, and Epstein, "Ground Systems as a Factor in Antenna Efficiency," *Proc. of the IRE*, Vol. 25, No. 6, June 1937.
2. Wait amd Pope, "Input Resistance of L.F. Unipole Aerials," *Wireless Engineer 32*, pages 131–138, May 1955.
3. Feldman, "The Optical Behavior of the Ground for Short Radio Waves," *Proc. of the IRE*, Vol. 21, No. 6, June 1933.
4. Jager, "Effect of the Earth's Surface on Antenna Patterns in the Short Wave Range," *Internet Elekitron Rundschau 24 (4)*, pages 101–104, 1970, Nr. 4.
5. King and Harrison, "The Impedance of Short, Long, and Capacitively Loaded Antennas with a Critical Discussion of the Antenna Problem," *Journal of Applied Physics*, Vol. 15, February 1944.
6. Friis, Feldman, and Sharpless, "The Determination of the Arrival of Short Radio Waves," *Proc. of the IRE*, Vol. 22, No. 1, January 1934.

Chapter 2

Short Ground-Mounted Verticals

C**hapter 1** reviewed the highlights of my first *QST* article, which was published in July 1971, and was entitled "The Ground-Image Vertical Antenna." From what I learned about ground-radials on the surface of the earth, I was able to construct a low-loss ground system for a parasitic vertical beam

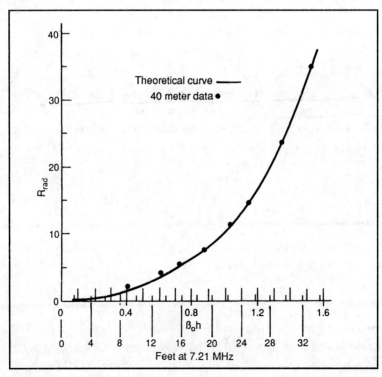

Figure 2-1. *Theoretical curve and experimental results for the radiation resistance as a function of height.*

(actually one-half of a Yagi beam) that proved to be very competitive. This antenna was described in the June 1972 issue of *QST* as "The W2FMI 20-Meter Vertical Beam."

In the process of trying to extend the results of these two investigations to the 40-, 80- and 160-meter bands, and to possible multiband use, the need arose to understand the operation of a shortened vertical and the effects of different loading schemes on the input impedance. I certainly was not interested in full-sized 1/4-wave verticals, especially on 80 and 160 meters. The results of this third investigation were published in the March 1973 issue of *QST* in an article entitled "The W2FMI Ground-Mounted Short Vertical." This chapter highlights some of the more important results presented in that article.

The first part of **Chapter 2** deals with the theoretical considerations of a short vertical antenna. This is followed by experimental results that show the trade-offs involved in shortening antennas by various schemes. Finally, specific designs are given for the 40- and 80-meter bands. As you will see, a short vertical antenna, properly designed and installed, approaches the efficiency of a full-size 1/4-wave vertical antenna. Even a 6-foot vertical on 40-meters can produce an exceptional signal. Moreover, efficient and broadband Ununs are now available to match 50-ohm cable to practically any short vertical antenna.

Theoretical Considerations

The short antenna has been defined as one that is small compared to a wavelength. In a more exact form, it is defined in such a manner as to simplify the mathematics in the theoretical calculations. King[7] has used the following inequality as the definition

$$\beta_0 h \leq 0.5 \qquad \text{(Eq 2-1)}$$

where:

$\beta_0 = 2\pi/\lambda$
h = half-length of a center-fed dipole or the height of a ground-mounted vertical
λ = the wavelength

$\beta_0 h$ is actually a quantity used to express the height of an antenna in terms of an angle in radians. Thus, this quantity is independent of the frequency. Because I used 40 meters most for the experimental results presented in this chapter, the inequality above assures accurate theoretical calculations for verticals of 11 feet or less.

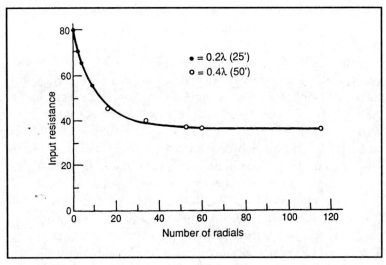

Figure 2-2. *Input resistance of a 1/4-wave ground-mounted resonant vertical as a function of the number of radials.*

The theoretical results show that the power gain (when compared to an isotropic radiator) for a *very* short antenna, even one that is less than 1 foot high on 40 meters, is 1.5.[7] This increases slowly to 1.513 for an 11-foot vertical. These gains are to be compared to about 1.62 for a resonant 1/4-wave vertical. This difference amounts to less than 0.4 dB or 0.07 S unit, based on 6 dB per S unit.

In a similar fashion, the related parameter known as the capture cross-section also differs by relatively small amounts. For the very short antenna, a cross-section value is $0.119\lambda^8$[8] while for the 1/4-wave vertical it is $0.13\lambda^8$. This is surprising to many because it is difficult to visualize that a vertical of a foot or two in height on 40 meters could have practically the same receiving ability as a 33-foot 1/4-wave antenna!

The *very small value of its input resistance* is the important property of a short vertical that makes its capture cross-section nearly the equivalent of a full 1/4-wave vertical. **Figure 2-1** shows the theoretical curve,[7] and the 40-meter experimental results for the input resistance of a ground-mounted vertical as a function of height. The experimental data were obtained by essentially canceling out the reactance of the shortened vertical by placing an inductance at its base and measuring the resistive value with a simple impedance bridge. Because I used an extensive ground system of 115 radials of No. 15 aluminum wire, each 0.4 wavelengths long, together with base loading coils with

Qs approaching 900 (see **Chapter 4**), the resistance measured was actually that of the antenna itself. This is called the *radiation resistance*.

Figure 2-2 shows the input resistance of a 1/4-wave resonant vertical as a function of the number of radials on the earth's surface and their termination in large nails (10 to 12 inches long). At the 115-radial point, the input resistance approaches the theoretical value of 35 ohms, which strongly indicates low earth loss and reliable data in short vertical antenna measurements. **Photo 1-A** in **Chapter 1** shows the experimental setup. Even though I used No.15 aluminum, I am quite sure that No. 22 or even No. 28 copper wire would work just as well.

It should also be noted from **Figure 2-2** that I noticed very little difference whether 0.2- or 0.4-wavelength radials were used. The results are practically identical with 16 radials. In fact, this

Photo 2-A. *Some of the different forms of loading that were used in this investigation.*

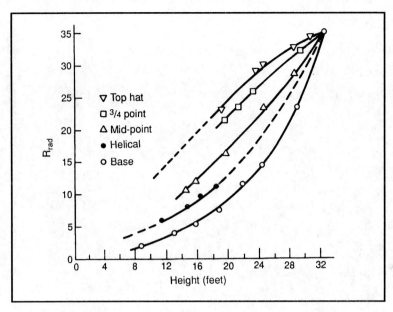

Figure 2-3. *Experimental results of radiation resistance as a function of height of antenna for various types of loading (40-meter data).*

curve is also similar to the one shown in **Chapter 1**. **Chapter 3** will show some results of the trade-offs in performance as a function of the lengths of the (unburied) radials.

Experimental Results

In designing a shortened vertical for 40, 80, or 160 meters, it is most important to obtain the highest possible resonant input impedance depending upon the mechanical constraints. It is a well-known fact that base-loading, shown in **Figure 2-1**, yields the lowest value. Because data on other forms of loading—i.e., top hat, three-quarter point, midpoint, and distributed (helical antenna)—were not readily available, I undertook an experimental investigation to see what increases could be obtained in the radiation resistance. **Photo 2-A** shows some of the different forms of loading used in this investigation.

The results of my experiments are shown in **Figure 2-3**. **Table 2-1** shows the individual points for inductive loading and **Table 2-2** for top hat and distributed loading. These experiments brought out several interesting results. According to **Figure 2-3**, top hat loading yielded the largest value of radiation resistance for a particular height. A top hat is also considered the most low-Q loading element. Surprisingly, the helical antenna[9] yielded a

1) Base Loading

No. of Turns*		h	R_{rad}
a)	7	24 feet 3 inches	14.5 ohms
b)	10	18 feet 9 inches	7.5 ohms
c)	12	15 feet 7 inches	5.5 ohms
d)	14	13 feet	4 ohms
e)	18	8 feet 10 inches	2 ohms

2) Midpoint Loading

No. of Turns*		h	R_{rad}
a)	6	28 feet 5 inches	28.5 ohms
b)	11	24 feet 3 inches	25.2 ohms
c)	15	19 feet 8 inches	16.5 ohms
d)	18	15 feet 8 inches	12.3 ohms
e)	24	14 feet 6 inches	10.5 ohms

3) Three-Quarter Point Loading

No. of turns*		h	R_{rad}
a)	10	29 feet 4 inches	32.5 ohms
b)	18	23 feet	26 ohms
c)	23	21 feet 2 inches	23.5 ohms
d)	24	19 feet 2 inches	22 ohms

*B&W 3029, 2-1/2 inch diameter, 6TPI, No. 12 wire.

Table 2-1. *Inductive loading.*

value less than midpoint loading. The three-quarter point and midpoint loading curves were not extended to lower values of height because data were very difficult to obtain below the points shown on the respective curves. The combinations of inductances and lengths below the heights shown on these two curves were probably beyond resonant conditions at the frequency used in the measurements. I extended the other curves with dashed lines indicating no difficulties were encountered in the measurement, and therefore other lengths were very possible.

Several other interesting comments can be made about top hat loading. **Table 2-2** shows that the reduction in height due to top hat loading (with a conducting wire around the perimeter) is approximately equal to twice the diameter of the top hat. Also, a four-spoked wheel approximates, to a good degree, a solid disk. Doubling to eight spokes only improves the loading by about 9 percent. Thus, *a few radials on the top of a vertical, which are electrically connected by a perimeter conductor, are very effective.* Four radials at the base, approximating a ground system, are practically *useless* as noted in **Figure 2-2**.

Top hats have been made successfully on 160 meters with sloping wires or struts.[8] Because of the sloping nature of the top hat, some canceling of the radiation occurs—reducing the radiation resistance further.

Although the curves in **Figure 2-3** were obtained from experiments at 7.21 MHz, these data can be applied to other bands by proper scaling. For example, by doubling all dimensions, including numbers of turns of the loading coils, the radiation resistance values would apply at a frequency of 3.6 MHz. By increasing the dimensions by 1.85 instead of 2, the results would apply at a frequency of 3.9 MHz. In like manner, a proper scaling factor could be used to apply these results to a portion of any of the amateur bands.

40- and 80-Meter Short Vertical Designs

The main objective in good short vertical antenna design is to have the radiation resistance much greater than the ohmic losses in the loading coils and ground systems. Coils with Qs approaching 900 (see **Chapter 4**) minimize the problem with loading coil losses. A sufficient number and length of ground-radials (for a ground-mounted vertical) eliminates the issue with ground losses. Because I used high-Q coils and 115 radials, any radiation resistance above a few ohms gave me the opportunity

1) Top Hat Loading (4-spoked wheel with 1/8-inch aluminum wire rim)

Diameter	h	R_{rad}
a) 1 foot	30 feet 10 inches	34 ohms
b) 2 feet	28 feet 7 inches	32.5 ohms
c) 4 feet	24 feet	30 ohms
d) 7 feet	19 feet 2 inches	23.5 ohms
e) 4 feet*	23 feet 4 inches	29.4 ohms

2) Distributed Loading (helical antenna)

No. of turns	h	R_{rad}	Top Hat at 1.5 ft. above Coil
a) 111	12 feet	8 ohms	1 foot dia.
b) 105	12 feet	10 ohms	2 feet dia.
c) 113	7 feet	6 ohms	2 feet dia.
d) 75	7 feet	7.5 ohms	4 feet dia.

*8-spoked wheel.

Table 2-2. *Top hat and distributed loading.*

Photo 2-B. *The author with the 6-foot, 40-meter vertical.*

to verify the theoretical predictions for very short verticals and to present designs that should be easily reproduced.

Because a broadband 4:1 Unun was available from my previous *QST* article on a 20-meter parasitic beam, the first design was for a short vertical having a resonant input impedance close to 12.5 ohms. The curves of **Figure 2-3** indicated that a 16- or 17-foot antenna with a coil of some 13 turns at 8 or 9 feet from the base would provide the proper impedance on 40 meters. I decided that some 7 to 8 feet of length above the coil could be replaced with a 4-foot top hat. The actual design that resulted had a total height of 10 feet and a 14-turn coil mounted 1 foot below the top hat. This height, was about 1 foot higher than first expected; but upon careful examination, I found that the top hat also lowered the radiation by replacing a section of the vertical portion. Therefore, the height had to be increased somewhat in order to counteract the top-loading effect.

I investigated several other shortened 40-meter vertical antennas. One was an 8.5-foot helical vertical using a 4-foot top hat and 75 turns of wire on a 1.625-inch, 7-foot-long wooden dowel. The input impedance was 7.5 ohms and it was matched to 50-ohm cable with a standard pi network. Several tests were made by doubling the winding pitch below and above the midpoint of the dowel, keeping the number of turns constant. There was very little difference in the results.

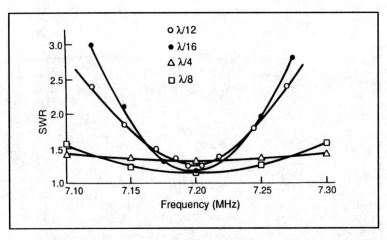

Figure 2-4. *Standing wave ratio of various short verticals compared to a 1/4-wave antenna. The 1/16-wavelength antenna is the 8-1/2-foot helical design, the 1/12-wavelength design is the 10-foot antenna, and the 1/8-wavelength one is the 15-foot, 10-inch antenna. All three are described in the text.*

The next design, using a 7-foot top hat and a 14-turn loading coil 6 inches below it, resulted in a resonant vertical only 6 feet high. The input impedance was only 3.5 ohms. Because an Unun with an impedance ratio of near 16:1 was not available at this time, matching was accomplished with a 4:1 Unun and a pi network. **Photo 2-B** (which was taken in 1972) shows the author and the 6-foot vertical.

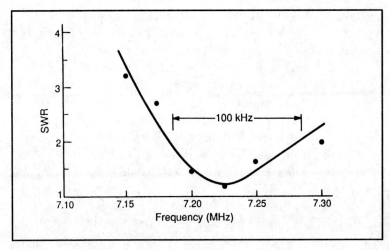

Figure 2-5. *Standing wave ratio of the 6-foot vertical using a 7-foot top hat and 14 turns of loading 6 inches below the top hat.*

Total Height	R_{rad} (ohms)	No. turns*	Diameter Top (feet)	Bandwidth
6 ft.	3.5	14 at 6 inches below Top Hat	7	100 kHz
8-1/2 ft.	7.5	75 on 7-foot dowel 1-8/5 inches in diameter	4	100 kHz
10 ft.	12.5	14 1 foot below Top Hat	4	125 kHz
15 ft., 10 in.	12.5	7 at base	4	540 kHz

*Except for helical antenna, coil wire is same as shown in Table 2-1.

Table 2-3. *Parameters of the 40-meter short vertical designs.*

I also investigated a larger 40-meter vertical. It had a 4-foot top hat and 7 turns of base loading that resulted in a height of 15 feet, 10 inches (approximately a 1/8 wavelength) and an input impedance of 12.5 ohms. The purpose of this design was to compare its low-angle radiation and bandwidth (defined as the range in frequency where the VSWR is less than 2:1) with the other antennas. **Table 2-3** shows the parameters of these various short verticals. **Figure 2-4** shows the VSWR curves of three of the short verticals compared to a 1/4-wave vertical and **Figure 2-5** shows the VSWR curve of the 6-foot antenna.

As the VSWR curves indicate, shortening an antenna generally decreases its bandwidth. Also, if one compares the 8.5-foot and 6-foot verticals, top hat loading appears to affect the bandwidth the least. Note that the 1/8-wavelength vertical appears to have a reasonable bandwidth. This could result in a practical design for a vertical beam that incorporates several 1/8-wavelength verticals.

Next, I compared the four short-vertical designs described above with a 1/4-wavelength vertical for low-angle radiation and performed on-the-air checks with other amateur signals. The low-angle radiation measurements were made three wavelengths in distance and at heights of 1, 3, 6, and 8 feet. This relates to vertical angles of 0.15, 0.45, 0.90, and 1.2 degrees. All measurements were made under matched conditions, with a constant 100 watts into the antenna under test. In no case were there any appreciable differences noted in the field strength measurements.

In fact, the 6-foot vertical seemed to give slightly higher readings. These measurements certainly tend to verify the theory on the power gain predicted for short verticals. On-the-air checks were again very gratifying and exciting. Over 200 contacts with the 6-foot antenna strongly indicated the efficiency and capability of a short vertical. Invariably, at distances greater than 500 to 600 miles, the short verticals yielded excellent signals.

After I had completed my investigation of short 40-meter verticals, I chanced upon a beach umbrella in a sports store. This umbrella had dimensions close to the 6-foot antenna used in my study. Without a second thought, I bought the umbrella and tried

Photo 2-C. *The author's XYL sitting under the W2FMI 40-meter beach umbrella antenna.*

it out over my low-loss ground system. **Photo 2-C** shows the final design shading my late XYL; **Figure 2-6** gives the parameters and VSWR curve. The antenna was matched to 50-ohm cable by a 16:1 Unun. Although the bandwidth was somewhat less than that of the 6-footer, the on-the-air results were about the same. This antenna was quite a conversation piece!

As I stated earlier, all of the results obtained on 40 meters can be obtained on the other bands with proper scaling. I tried this

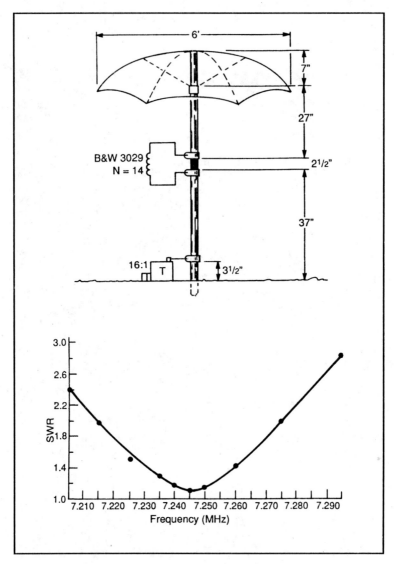

Figure 2-6. *The parameters and standing wave ratio as a function of frequency for the W2FMI 40-meter beach umbrella antenna.*

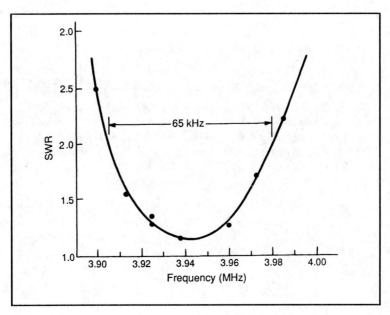

Figure 2-7. *The standing wave ratio for the 22-foot 80-meter vertical.*

out on 80 meters. Because a 7-foot top hat (rather than an 8-foot model) was available, the height actually turned out to be 22 feet instead of 20. The loading coil had 24 turns and was placed 2 feet below the top hat. The bandwidth was 65 kHz (half of that of the 40-meter 10-footer) as is shown on **Figure 2-7**.

Conclusions

Several interesting results came out of this investigation which, even to one schooled in antenna theory, are difficult to believe. I refer specifically to the ability of very short verticals to radiate and receive as well as a full-size 1/4-wave antenna. The differences are practically negligible; but, the trade-offs are in lowered input impedances and bandwidths. However, with a good image plane and proper design, these trade-offs can be entirely acceptable.

Another point should be mentioned in relation to the results reported here. Even though my results were obtained using short ground-mounted verticals, they are valid for center-fed verticals or horizontal antennas, as well. The only difference is that one must double the impedance values and account for the effect of the image antenna.

I would like to acknowledge the support of many amateurs for their fine words of encouragement and excellent reporting during the on-the-air contacts. During the course of this study, some 350 amateurs reported on comparisons between signals from these short verticals and those of other stations.

References

7. King, *The Theory of Linear Antennas*, Harvard University Press, Cambridge, MA 1956.
8. Weeks, *Antenna Engineering*, McGraw-Hill Book Company, 1968.
9. The diameter of the helical antenna was 1-5/8 inches. Each helical antenna had a top hat 18 inches above the helix in order to terminate the upper turns with sufficient capacitive reactance.

Chapter 3

Short Ground-Radial Systems for Short Verticals

Again, **Chapter 1** reviewed my July 1971 *QST* article. This article, which investigated radials on the earth's surface, showed that four 1/4-wave radials make a *very* poor ground system. With such a system, practically half of the output power is lost to the earth. It takes about 40 1/4-wave radials to reduce the loss to only a few ohms. Furthermore, the low-angle radiation is improved by the addition of more and longer radials. The article also showed that radials need not be buried to be effective. Because the electric field does not penetrate the earth's surface deeply at HF, the optimum radial depth is probably one which lets you cut the lawn safely, and also prevents the radials from becoming a tripping hazard to those using the yard. Additionally, it was noted that practically any size and type of conducting wire can be used to create a satisfactory ground-radial system.

Chapter 2 reviewed my March 1973 *QST* article, which dealt with the theory and practice of short, ground-mounted verticals. It investigated all the usual types of loading schemes and presented various short vertical antenna designs. Using measurements on verticals over an extensive ground system of 115, 0.4 wavelength radials, it revealed that a properly designed short vertical approaches the performance of a full-size 1/4-wave vertical. The only real compromise is in bandwidth. Interestingly enough, my investigation also revealed that radials only 0.2-wavelengths long performed as well as the much larger 0.4-wavelength radials.

The major questions I had after writing these two articles were: 1) How short can the radials be and still yield acceptable results?; and 2) What is the soil conductivity under and in the near vicinity of the verticals I used? **Chapter 3** reviews my

Material	Conductivity (millimhos/meter)
Poor soil	1–5
Average soil	10–15
Very good soil	100
Salt water	5000
Fresh water	10–15

Table 3-1. *General classification in conductivity.*

April 1978 article, "Short Ground-Radial Systems of Short Verticals," which presents experimental evidence that answers these questions. My investigation showed that very small radial systems made up of almost any type of thin wire placed on the earth's surface can perform surprisingly well. On-the-air comparisons with a much larger vertical over an extensive ground system showed performances reduced by only a few decibels. Also, a new soil conductivity measurement technique, introduced for the first time in this article, indicated that the conductivity at my location is average to a little above average. Incidentally, several years after this article appeared in *QST*, I was informed that the soil conductivity technique presented in my article had been selected as the world standard.

Introduction

Vertical antennas have enjoyed considerable success on the 80- and 160-meter bands due to the difficulty of erecting horizontal antennas at heights sufficient for low-angle radiation. Optimum heights, which are in access of 1/2 wavelength on these bands, are impractical for most amateurs. Often short verticals are used, because they have been shown to compare favorably with full-size 1/4-wave verticals. This is true if losses in the ground system, matching networks, and loading elements are small compared to the radiation resistance of the short vertical.

Considerable information is available that describes the effects of buried radials on the efficiency of 1/4-wave verticals in the MF and LF bands as a function of the length and number of radials, and the conductivity of the soil.[1,2,4,10-23] However, there is little information on radials lying on the ground's surface—particularly in connection with short verticals.

Soil Conductivity

The conductivity of the soil under and in the near vicinity of a ground-mounted vertical is important in determining the extent of the radial system required and the overall performance that will be achieved. Most soils are nonconductors of electricity when completely dry. Conduction through the soil results from conduction through the water held in the soil. Thus, conduction is electrolytic. DC techniques for measuring conductivity are impractical because they tend to deplete the carriers of electricity in the vicinity of the electrodes. The main factors contributing to the conductivity of the soil are:

1) Type of soil.
2) Type of salts contained in the water.
3) Concentration of salts dissolved in the contained water.
4) Moisture content.
5) Grain size and distribution of material.
6) Temperature.
7) Packing density and pressure.

Although the type of soil is an important factor in determining its conductivity, rather large variations can take place from one location to another because of the other factors involved. Generally, loams and garden soils have the highest conductivities. These are followed, in order, by clays, sand, and gravel. Soils have been classified as shown in **Table 3-1**. Although some differences are noted in the reporting[24,25] of this mode of classi-

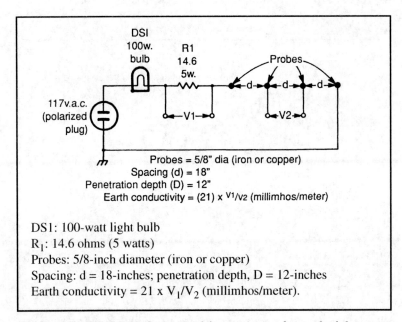

Figure 3-1. *Schematic diagram of four-point probe method for measuring earth conductivity.*

fication because of the many variables involved, the classification generally follows the values shown in the table.

Because conduction through the soil is almost entirely electrolytic, AC measurement techniques are preferred. Many commercial instruments that use AC techniques are available and described in the literature.[26] However, there are rather simple AC measurement techniques which provide accuracies on the order of 25 percent and are quite adequate for the radio amateur. Such a setup was developed by a Bell Labs colleague and neighbor, M.C. Waltz,[27] W2FNQ. It is shown schematically in **Figure 3-1**. **Figure 3-2** shows the conductivity readings taken over the last three months of 1976. It is interesting to note the general drop in conductivity over the three months studied, as well as the short-term changes due to periods of rain. The results presented in the following sections on antenna efficiencies were obtained during the period from October 10 to November 10, 1976, when the conductivity varied between 22 and 25 millimhos/meter. As **Table 3-1** shows, my soil can be classified as average to slightly above average in conductivity.

Antenna Efficiency Considerations

The antenna efficiencies I will describe are based upon the losses that appear in series with the radiation resistance of resonant verticals. Although this approach does not give a comparison

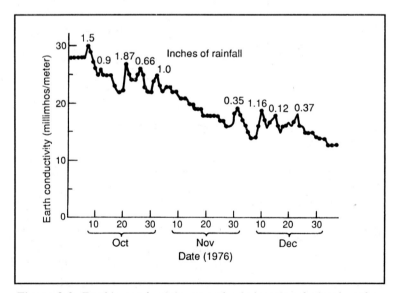

Figure 3-2. *Earth's conductivity at author's location during last three months in 1976.*

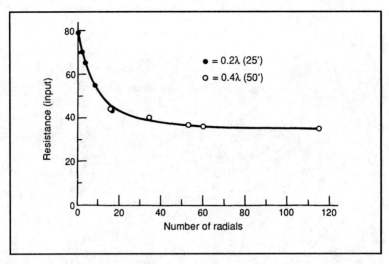

Figure 3-3. *Input impedance of resonant 1/4-wave vertical as a function of the number of radials.*

between the very low angles of radiation (i.e., less than 15 degrees) of various radial systems, it does allow for comparisons in the 15- to 30-degree range—which is important for sky-wave transmission on the 40-, 80-, and 160-meter bands. Mathematically, this definition for antenna efficiency can be written as:

$$\text{Antenna efficiency} = \frac{R_{rad}}{R_{rad} + R_g + R_A} \qquad \textbf{(Eq 3-1)}$$

where:

R_{rad} = radiation resistance
R_g = ground loss
R_A = ohmic losses due to loading and the antenna itself.

With high-Q loading coils and practically any size of aluminum tubing for the antenna, R_A can be minimized and, therefore, eliminated from the relationship above.

An example of this technique for determining antenna efficiency uses the results shown in **Figure 3-3**. The input impedance of a resonant 1/4-wave vertical is plotted as a function of the number of radials. Two lengths of radials (0.2- and 0.4-wavelengths) were considered. Because the radiation resistance for the thickness of the vertical used in this experiment is 35 ohms, the losses with 50 radials were approximately 2 ohms. With 100 radials, losses were about 1 ohm. This amounts to efficiencies of

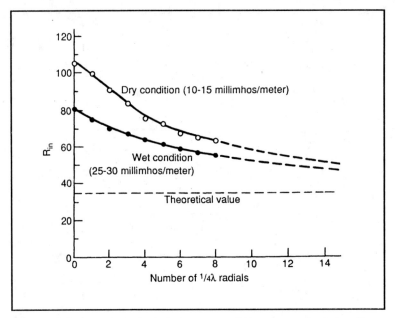

Figure 3-4. *Input impedance of resonant 1/4-wave vertical as a function of the number of radials and the condition of the soil.*

94 and 97 percent, respectively. The efficiency with only four radials is less than 60 percent. This poor efficiency level exists even for a location with a soil conductivity that can be considered average to a little above average.

Furthermore, the efficiency of a radial system employing small numbers of radials is quite dependent on the moisture content of the soil. **Figure 3-4** shows what happens with a resonant 1/4-wave vertical on 20 meters, as the number of radials varies from one to eight. The difference in efficiency between wet and dry conditions becomes less pronounced as the number of radials is increased. The efficiency also becomes more independent of soil conductivity as the number of radials is increased.

To determine the efficiencies of shortened verticals over abbreviated radial systems, I compared resonant input impedances with similar antennas over a near-ideal radial system. **Figure 3-5** shows the experimental results for the various kinds of loading, as a function of height, over a near-ideal image plane (115 radials on the ground, about 50 feet long, and terminated in 10- to 12-inch nails). These results, which have been most useful to me in designing short verticals, are described in more detail in **Chapter 2**.

Because all of my radials were on the surface of the earth and terminated in large nails, I undertook a study to determine the effect of the length of the spike. I measured the efficiency for resonant 1/4-wave verticals with small numbers of shortened radials on the 20- and 40-meter bands. The radials were four and eight feet long (1/16th wavelength), respectively. **Figure 3-6** provides the results for four different depths of termination. The figure shows that depths of 10 to 12 inches should be sufficient for the soil conductivity at my location for 20 and 40 meters, and most likely for 80 and 160 meters as well. Incidentally, the effectiveness of the radials of a 1/4-wavelength or longer did not change noticeably as a function of the depth of the termination. Therefore, terminations for radials about 0.2-wavelengths and longer are primarily used for mechanical reasons.

Abbreviated Radial Systems

In order to determine experimentally the efficiency of shortened verticals with abbreviated radials on the earth's surface, I used five verticals of different heights and loading schemes on the 20- and 40-meter bands. My results were then compared with similar antennas over a near-ideal image plane as shown in **Figure 3-5**. The five resonant verticals selected were:

1) 1/4 wavelength
2) 3/16 wavelength, top hat loaded

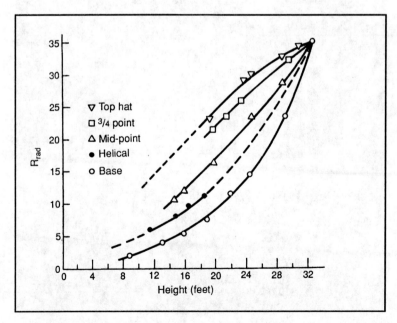

Figure 3-5. *Experimental results of radiation resistance as a function of height of antenna for various methods of loading.*

3) 3/16 wavelength, midpoint loaded
4) 1/8 wavelength, top hat loaded
5) 1/8 wavelength, midpoint loaded

Chapter 2 presents some of the results for these 10 verticals.

My test range used a radial system with various lengths of No. 17 steel fence wire terminated with 10- to 12-inch nails. **Photo 3-A** shows the 12-inch aluminum base plate and the input connection arrangement. This antenna system was erected in the front yard approximately 50 feet from the house, and offered an opportunity for on-the-air comparisons with verticals mounted on the near-ideal ground system in the backyard.

The results are shown in **Figures 3-7** and **3-8**. Only the 40-meter data is presented, as little difference was noted on 20 meters. **Figure 3-7** shows the effect on the efficiency of a resonant 1/4-wave vertical on 40 meters as a function of the number of radials using three different lengths of terminated radials. Although these curves were obtained on a 40-meter system, the

Photo 3-A. *The 12-inch aluminum plate and input connection arrangement. Shown are 48 radials of No. 17 steel electric fence wire.*

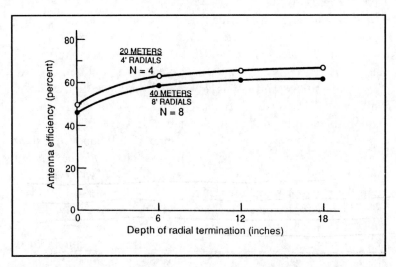

Figure 3-6. *Efficiency of resonant 1/4-wave verticals on 20 and 40 meters as a function of the length of spike terminating the radials.*

relationships are generally valid for all other frequencies in the HF range if the same fractional wavelengths are used for the radials. As the figures show, the longer 1/4-wave radials yielded the highest efficiency. However, it is interesting to note that this improvement in efficiency with length decreases as the number of radials decreases. At four radials, the efficiency with 8-foot

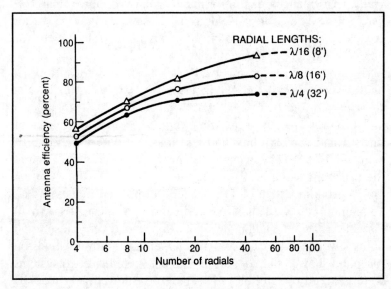

Figure 3-7. *Efficiency of resonant 1/4-wave vertical as a function of the number of terminated radials with three different lengths.*

31

radials is not much poorer than with 32-foot radials; i.e., 50 percent compared to 56 percent. Other interesting trade-offs also exist with various lengths and numbers of radials. **Figure 3-7** shows that 16 1/4-wave radials are about equivalent to thirty-five 1/8-wave radials, and eight 1/4-wave radials are equivalent to only twelve 1/16-wave radials. Obviously, other equivalencies can be obtained from the figure.

Figure 3-8 shows the results of antenna efficiency for short verticals as a function of the length of radials, with the number of radials kept constant at 48. As expected, 1/4-wave verticals, with their higher radiation resistance, have the highest efficiencies. Surprisingly, the efficiency of the 1/8-wave vertical does not suffer proportionally. That is, the 1/8-wave vertical with midpoint loading still has a 67-percent efficiency compared to the 84-percent efficiency of the 1/4-wave vertical, even though its radiation resistance is only one-third as large (12.5 compared to 35 ohms). A further comparison with a 1/4-wave vertical over an ideal ground system predicts that the 1/8-wave vertical with 48 1/8-wave terminated radials should show a reduced performance of only 1.7 dB.

On-the-Air Comparisons

As was shown in the preceding section, short verticals with a sufficient number of abbreviated and terminated radials should yield performances of only a decibel or two poorer than 1/4-wave verticals over near-ideal ground systems. Many on-the-air comparisons were made to confirm this prediction, which was based on efficiency considerations.

The first comparison involved a 40-meter, 1/8-wave, top hat loaded vertical with forty-eight 1/8-wave (16-foot) radials (No. 17 steel wire on the ground and terminated with 10- to 12-inch nails). It was erected in the front yard. The input impedance of this vertical was 25 ohms and was matched with a highly efficient 2:1 Unun. This antenna system was compared with one in the backyard using a 29-foot vertical with a 13.5-foot top hat over a ground system of 115, 50-foot terminated radials of No. 15 aluminum wire. This larger vertical was resonated by a small variable capacitor yielding an input impedance very close to 50 ohms. Over 100 contacts on 40 meters plus 200 observations on reception showed that the differences between these two systems were generally negligible. A few reports showed a 1- to 2-dB difference in favor of the much larger system, but these were in the minority.

An even more interesting comparison was made on 80 meters. A 20-foot vertical with an 8-foot top hat was erected in the front yard over the same ground system of forty-eight 16-foot, No. 17 steel wire radials (terminated). A base-loading coil of about 20 turns (6 tpi) of No. 12 wire with a diameter of 2.5-inches was required to resonate the antenna. The input impedance was close to 12 ohms. A 4:1 Unun was used to match it to 50-ohm cable. This antenna system represents a radiation resistance of about 5 ohms and a ground loss of about 7 ohms. The high-Q loading coil had virtually no appreciable loss. This antenna was compared to the vertical in the backyard, which was 29 feet tall and had a 13.5-foot top hat and was over a near-ideal ground system. This taller antenna required about eight turns of No. 14 wire on a powdered-iron core (T200-2) to resonate on 80 meters. Its input impedance was 15 ohms (showing negligible loss in the ground system and loading coil), and matching was accomplished with an efficient 3.33:1 fractional-ratio Unun. Again, about 100 comparisons were made on the air and another 200 observations were made on reception. The average difference between the two systems amounted to only about 5 dB in favor of the much larger system in the back yard. This is quite noteworthy as many contacts with the larger system established it as a very competitive antenna system.

Concluding Remarks

Quarter-wave verticals over an extensive radial system have been known to be efficient, low-angle radiators. Even short verticals over the same large ground system have been shown to lose little in the way of performance. With low-loss Ununs and matching techniques, short verticals suffer only in bandwidth. However, full-sized 1/4-wave verticals and near-ideal ground systems are beyond the reach of most radio amateurs on 80 and 160 meters. This investigation was undertaken because little information was available on limited radial systems, particularly for short verticals.

As was shown, short radials over soil of average to a little above average conductivity can perform quite acceptably for verticals of all heights. The results of this investigation now let one predict quite accurately the operation of verticals less than a 1/4-wave and with radials as short as 1/16-wave. The simple soil-conductivity measurement technique described also provides a tool for comparing a given location with others, as well as predicting the performance of a ground-mounted vertical antenna.

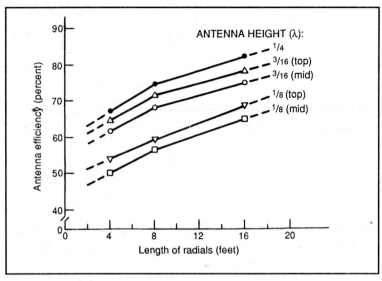

Figure 3-8. *Efficiency of short verticals as a function of the length of the terminated radials, with the number kept constant at 48.*

Multiband vertical antennas, which have recently become commercially available, have demonstrated that the radial systems described here are not required because they operate electrically as 1/2-wave antennas. This is very likely true. Because it's nearly impossible to use this principle on 80- and 160-meters, the shortened vertical with abbreviated radials still appears to be the solution for low-angle radiation for many, if not most, radio amateurs.

References

10. Abbott, "Designs of Optimum Buried RF Ground Systems," *Proc. IRE*, pages 846–852, July 1952.
11. Wait and Pope, "The Characterization of a Vertical Antenna with a Radial Conductor Ground System," *Appl. Sci. Research* (B), pages 177–195, 1954.
12. Monteath, "The Effect of the Ground Constants and an Earth System on the Performance of a Vertical Medium-Wave Aerial," *Proc. IRE*, Vol. 105C, Pt 2, pages 292–306, January 1958.
13. Smith and Devaney, "Fields in Electrically Short Ground Systems: An experimental Study," *J. Res. NBS* 63D Radio Prop. No. 2, pages 175–180, September/October 1959.
14. Larsen, "The E-Field and H-Field Losses Around Antennas with a Radial System," *J. Res. NBS* 66D Radio Prop. No. 2, pages 189–204, March/April 1962.
15. Maley and King, "Impedance of a Monopole Antenna with a

Circular Conducting-Disk Ground System on the Surface of a Lossy Half Space," *J. Res. NBS* 65D Radio Prop. No. 2, March/April 1961.

16. Maley and King, "Impedance of a Monopole Antenna with a Radial-Wire Ground System on an Imperfectly Conducting Half-Space, Part I," *J. Res. NBS* 66D Radio Prop. No. 2, March/April 1962.

17. Wait and Walters, "Influence of a Sector Ground Screen on the Field of a Vertical Antenna," *NBS* Monograph 60, April 15, 1963.

18. Maley and King, "Impedance of a Monopole Antenna with a Radial-Wire Ground System on an Imperfectly Conducting Half-Space, Part II," *J. Res. NBS* USNC-URSI, 68D, No. 2, February 1964.

19. Maley and King, "Impedance of a Monopole Antenna with a Radial-Wire Ground System on an Imperfectly Conducting Half-Space, Part III, *J. Res NBS* USBC-URSI, 68D, No. 3, March 1964.

20. Gustafson, Chase, and Balli, "Ground System Effect on High Frequency Antenna Propagation," *Research Report*, U.S. Navy Electronics Laboratory, January 4, 1966.

21. Hill and Wait, "Calculated Pattern of a Vertical Antenna with a Finite Radial-Wire Ground System," *Radio Science*, Vol. 8, No. 1, pages 81–86, January 1973.

22. Rafuse and Ruze, "Low Angle Radiation from Vertically Polarized Antenna Over Radially Heterogeneous Flat Ground," *Radio Science*, Vol. 10, pages 1011-1018, December 1975.

23. Stanley, "Optimum Ground Systems for Vertical Antennas," *QST*, December 1976.

24. Card, "Earth Resistivity and Geological Structure," *Electrical Engineering*, pages 1153–1161, November 1935.

25. *Reference Data for Radio Engineers*, Fifth Edition, Howard W. Sams and Co., Inc., ITT, pages 26-3 to 26-5.

26. Lagg, G.F., *Earth Resistances*, Pitman Publishing Corp., 1964, pages 206–229.

27. Private communication.

Chapter 4

The Loading Coil

In 1989, after completing work on the second edition of my book *Transmission Line Transformers*, I thought it would be interesting to look at short vertical antennas for mobile use in the HF band and ground planes made from steel fencing and screening for field day use. I had purchased 20-, 40-, and 80-meter low-power (400-watt) Hustler whip antennas 10 years before, and had always intended to equip my car for mobile operation. Now, I finally had the opportunity to characterize these whip antennas over a large ground system of eight radials of steel fencing (with 2-inch x 3-inch openings) three feet wide and 25 feet long, resulting in a ground loss of only 2 to 3 ohms on the three bands. Then I acquired the three Hustler high-power (2-kW) resonators to see what improvements they would make in the antennas' efficiencies. To my surprise, the differences in the input impedances of the 20- and 40-meter antennas were just about the same with either the low- or high-power resonators. This meant that the losses were comparable. However, there was a large difference in the losses between the 80-meter resonators. The 2-kW resonator had an input impedance of 35 ohms, while the 400-watt resonator had only 21.5 ohms. This indicated that the high-power loading coil could have about 15 ohms more loss! To better understand what the input impedance measurements meant, I decided to try to obtain a Q meter and study the fundamentals of the loading coil itself.

I borrowed a Boonton 260-A Q meter from a local amateur and began a characterization of the Hustler resonators and many other loading coils. Because the Boonton Q meter was an older (circa 1940) vacuum-tube model, accurate readings were difficult to obtain due to the necessary corrections needed in the Q and frequency readings. Luckily, my colleagues at AT&T Bell Laboratories had a later (circa 1970) solid-state model, a Hewlett-Packard 4342A Q meter, that I was able to borrow for a few weeks. This model didn't require me to make any corrections to the readings and, therefore, the readings were not only more accurate, but were also obtained in much less time.

Just as I was about to publish the results of my investigation, I received considerable correspondence containing critical feedback from readers regarding the second edition of my book. These concerns all required immediate attention. The readers were all having problems obtaining the components for my Balun and Unun designs. Most of them were not available retail! By working closely with Amidon Associates, I was able to redesign practically all of the transformers in my book with components that were readily accessible. In the process of redesigning many of the transformers, I came across new terms in the literature like *voltage*, *current*, and *choke* Baluns—with which I was not familiar. After a more serious review of the amateur literature, I decided to not only publish my new designs, but my views on these new terms as well. The results of my new designs and views are the basis for this book.

Because my work on transmission line transformers is now complete (at least for the time being), I thought it might be interesting to present some of the highlights of my investigation on the loading coil. Obviously, what I've learned is relevant to the preceding three chapters.

I'll begin by discussing the theoretical considerations with loading coils and Q meters. I'll follow this with experimental results that include the Hustler resonators, B&W* coils, homemade coils and the effects of the mounting hardware. As you will see, the modeling technique I have used indicates that high-Q coils can reach values in the 800 to 900 range. These values are generally two to three times greater than expected. They are also two to three time greater than the values that have appeared in the literature.

Theoretical Considerations

Figure 4-1A shows the actual schematic of a coil and **Figure 4-1B** its equivalent series schematic. In **Figure 4-1A**, R_{ac} equals the AC coil resistance (which varies with frequency), L_0 equals the low-frequency inductance (where C_d plays no role), and C_d equals the distributed capacitance (the capacitance between different parts of the coil).

*Barker & Williamson, Inc., 10 Canal Street, Bristol, Pennsylvania 19007.

Applying the usual circuit analysis to the parallel network of **Figure 4-1A**, we have for the impedance:

$$Z = \frac{R_{ac} + J\omega [L_0 (1-\omega^2 L_0 C_d) - C_d R_{ac}^2]}{(1-\omega^2 L_0 C_d)^2 + \omega^2 C_d^2 R_{ac}^2} \qquad \text{(Eq 4-1)}$$

where:

$$\omega = 2\pi f$$

At frequencies that do not exceed 80 percent of the self-resonant frequency of the coil, **Equation 4-1** reduces to:

$$Z = \frac{R_{ac}}{(1-\gamma^2)^2} + \frac{J\omega L_0}{(1-\gamma^2)} \qquad \text{(Eq 4-2)}$$

where:

$$\gamma^2 = \left(\frac{f}{f_0}\right)^2 \qquad \text{(Eq 4-3)}$$

and

$$f_0 = \text{self-resonant frequency} = \frac{1}{2\pi} \sqrt{\frac{1}{L_0 C_d}} \qquad \text{(Eq 4-4)}$$

Thus, from the equivalent circuit of **Figure 4-1B**, we see that:

$$R_e = \frac{R_{ac}}{(1-\gamma^2)^2} \qquad \text{(Eq 4-5)}$$

and

$$L_e = \frac{L_0}{(1-\gamma^2)} \qquad \text{(Eq 4-6)}$$

and

$$\frac{\omega L_e}{R_e} = \text{equivalent Q of a coil} = Q_e \qquad \text{(Eq 4-7)}$$

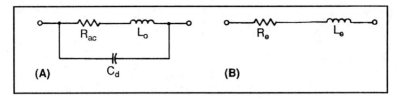

Figure 4-1. *(A) is the actual schematic of a coil with distributed capacitance and (B) is the equivalent series schematic.*

The Q meter actually uses the equivalent circuit of **Figure 4-1B** in a series arrangement to bring about series resonance. The series equivalent circuit is also most easily visualized in loading-coil applications. **Figure 4-2** shows the series resonant circuit of a Q meter.

At resonance:

$$X_C = X_L \quad \text{(Eq 4-8)}$$

and

$$i = e/R_e \quad \text{(Eq 4-9)}$$

Therefore, the voltage across the variable capacitor at resonance becomes:

$$E = e/R_e \cdot X_C \quad \text{(Eq 4-10)}$$

and

$$X_C/R_e = E/e = Q_e \quad \text{(Eq 4-11)}$$

Equation 4-11 is correct for values of Q equal to or greater than 10. Therefore, if e is held to a constant and known level, a

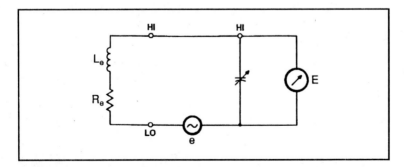

Figure 4-2. *The Q meter series resonant schematic.*

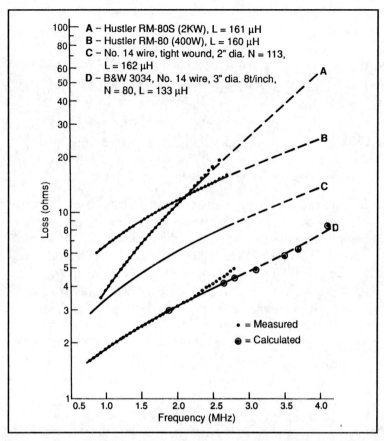

Figure 4-3. *A comparison of losses with four different 80-meter loading coils used with whip antennas.*

voltmeter can be connected across the capacitor and calibrated directly in terms of Q.

However, there is still one important problem with Q meters. They cannot measure coils with very high Qs. Hewlett-Packard acknowledged this problem and said that they stop measuring coils when the Qs exceed 500. Therefore, to obtain some acceptable values for high-Q coils, I resorted to a curve-fitting technique for R_{ac}, which is the most difficult component to obtain. L_o and low-frequency values of R_{ac} are readily obtainable with the Q meter. The self-resonance and hence C_d are readily obtained with a gate-dipper and a modern transceiver.

The success in modeling the coil now rests upon the determination of R_{ac} with frequency. A good start is to assume that this parameter varies as the square-root of frequency as a result of *skin effect*.[28] Knowing accurate low-frequency values of R_{ac} and multiplying these values by the square-root of the frequency ratio yields:

$$R_{ac}|_{f_2} = R_{ac}|_{f_1} \left(\frac{f_2}{f_1}\right)^{1/2} \quad \text{(Eq 4-12)}$$

where:

f_1 = the lower frequency where an accurate value of R_{ac} is obtained

f_2 = the upper frequency where R_{ac} is calculated

By calculating several points within the region where accurate measurements are known to be achieved, one is able to determine if the calculated points are a good extension of the low frequency measurements. The square-root dependency works well for coils that have their turns spaced one wire diameter. For a tightly wound coil that has a considerable *proximity effect*[29] (crowding the current to the inner and outer surfaces of the con-

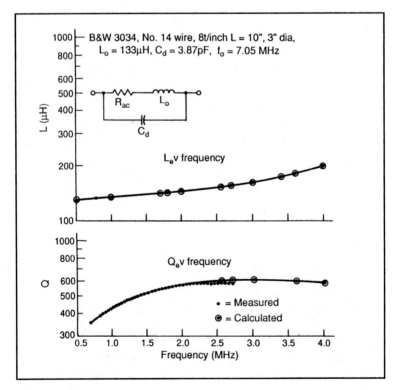

Figure 4-4. *The variations in the equivalent inductance, L_e, and Q_e with frequency of the low-loss loading coil, B&W 3034, used in the 80-meter whip in Figure 4-3.*

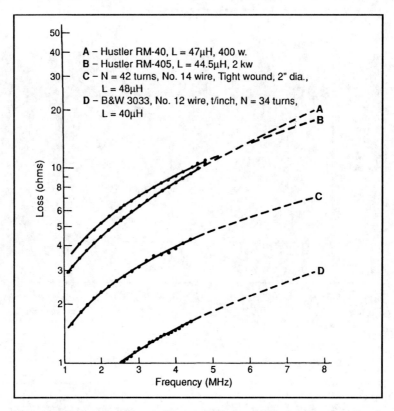

Figure 4-5. *A comparison of losses with four different 40-meter loading coils used with whip antennas.*

ductors), the frequency dependency has been found to vary as the square of the cube-root of the frequency ratio, that is:

$$R_{ac}|_{f_2} = R_{ac}|_{f_1} \left(\frac{f_2}{f_1}\right)^{2/3} \qquad \textbf{(Eq 4-13)}$$

Thus, with hand calculators that have the function y^x, one is able to determine quite precisely what fractional power to raise the frequency ratio for a better fit to the curves.

Experimental Results

The first measurements and comparisons were made on the 80-meter loading coils because I wanted to find out why the Hustler high-power whip antenna had a much higher input impedance over a low-loss ground system. The answer lies in **Figure 4-3**. At 3.5 MHz, the high-power resonator (loading coil) displayed a loss about twice that of the low-power unit—40 ohms compared to 20

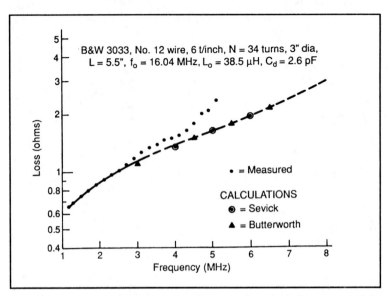

Figure 4-6. *A later look at the variation in loss with frequency of the low-loss loading coil, B&W 3033, used in the 40-meter whip shown in Figure 4-5.*

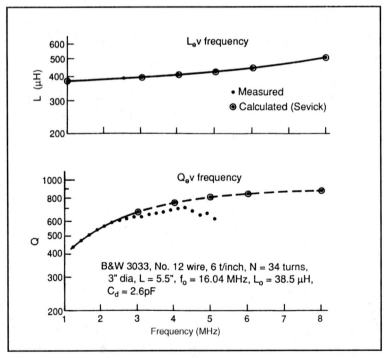

Figure 4-7. *The variations in equivalent inductance and Q with frequency for the low-loss coil, B&W 3033, shown in Figure 4-6.*

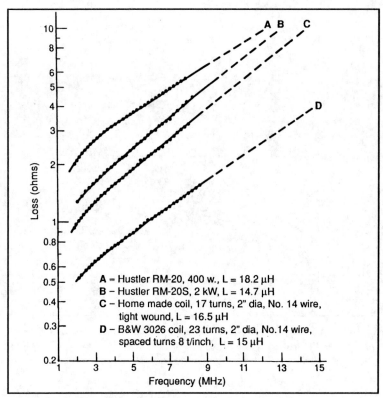

Figure 4-8. *A comparison of losses with four different 20-meter loading coils used with whip antennas.*

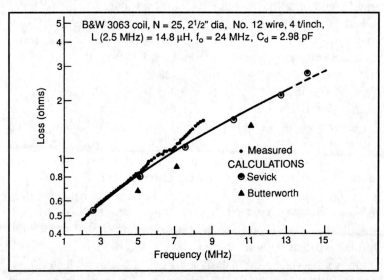

Figure 4-9. *A later look at the variation in loss with frequency of another low-loss loading coil, the B&W 3063, used with the 20-meter whip. It has a little thicker wire (No. 12) and larger diameter than the one shown in Figure 4-8.*

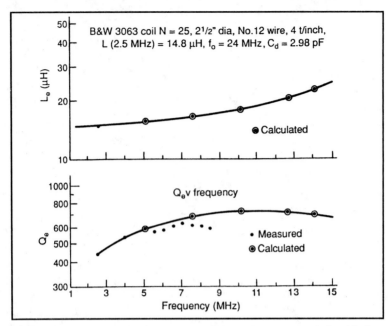

Figure 4-10. *The variations in equivalent inductance and Q with frequency for the low-loss coil, B&W 3063, shown in Figure 4-9.*

ohms. Even though the RM-80-S resonator used a tightly wound, No. 14 Litz wire coil, its main loss resulted from using metal end-caps very close to the ends of the coil (less than 1/4 inch). As such, they not only induced excessive loss, but also significantly reduced the inductance of the coil as well. Even though the 400-watt unit used No. 20 wire wound tightly on a smaller coil form, its metal end-caps were considerably further from the ends of the coil, thus accounting for the higher efficiency.

Curve C shows what can be done with No. 14 H Thermaleze wound tightly on a 2-inch diameter plastic form (from the grocery store) with no end-cap effect. Finally, curve D shows what can be achieved with a high-Q coil that has a spacing of about one wire-diameter between the turns. **Figure 4-4** shows the variation in inductance and Q with frequency for the B&W, 80-meter loading coil. Even though the inductance was less than that of the other three (the coil is only obtainable in 10-inch lengths) and, therefore, requires a little extra length above the coil for resonance, its improved efficiency is obvious.

Figure 4-5 shows some early comparisons with 40-meter loading coils. In this case, the low-power and high-power resonators show about the same loss at 7 MHz. Again, curve C shows what can be achieved by eliminating the end-cap effect. Curve D shows what happens when the turns are spaced. **Figures 4-6** and **4-7** provide a later view of the B&W 40-meter loading coil. They also include some computations from Butterworth's work.[29] Note the disparity between the experimental and calculated results beyond 3 MHz where the Q is greater than 600!

Figure 4-8 shows early comparisons with the 20-meter loading coils. The high-power Hustler units use No. 14 spaced wire on a 2-inch diameter form. The low-power unit has many No. 20 close-wound turns on a 1/2-inch diameter coil form. Curve C shows what can be done by eliminating the end-cap effect and curve D, and also the effect of spacing the turns. **Figures 4-9** and **4-10** show what can be obtained with a coil using a slightly thicker wire and larger diameter.

Figures 4-11 and **4-12** show what can be obtained with 40-meter loading coils for mobile applications using No.16 wire.

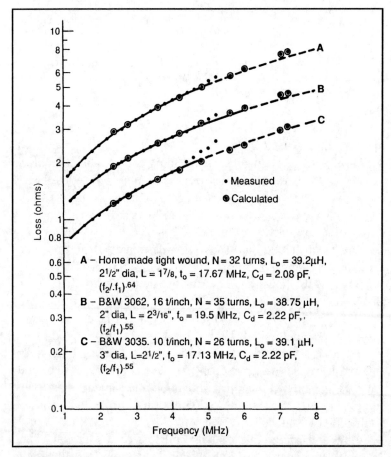

Figure 4-11. *A comparison in the losses versus frequency of three loading coils using No. 16 wire with different spacings between turns (primarily).*

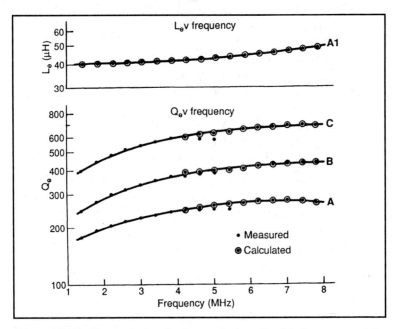

Figure 4-12. *The variation in the equivalent Q with frequency of the three coils of Figure 4-11 and the variation in inductance with frequency for the tightly wound coil.*

Curve C in **Figure 4-12**, the B&W 3035 coil, shows a Q approaching 700 at 8 MHz!

Figure 4-13 shows the loss in a loading coil as a function of the spacing of the hardware in a concentric mounting. The loss nearly triples when the spacing goes from zero to infinity (only the nylon rod itself). **Figure 4-14** shows the loss versus frequency when a coil similar to the one in **Figure 4-13** is mounted 1.5 inches from the aluminum tubing and insulator (side-mounted). In this situation, the loss increases from 2.3 to 2.8 ohms at 3.5 MHz.

Figure 4-15 shows the experimental results of the sensitivity of loss to the spacing of metal end-caps. This experiment was prompted by the large loss exhibited by the Hustler high-power 80-meter resonator. At 3.5 MHz, the loss doubles in going from zero spacing to infinity (no end-caps at all). Although this information is not shown, I also investigated copper and steel end-caps. The results with copper were about the same as that of aluminum. However, steel showed about three times as much loss as the other two metals.

Figure 4-16 attempts to show coil loss as a function of the length-to-diameter ratio, l/D, of coils with relatively the same inductance. Because considerable space has been devoted to this subject in the literature, I was surprised at the small difference observed in going from an l/D of 3.16 to 1.17. At 3.5 MHz, it shows an improvement of only 0.5 ohms. Even though the ideal l/D ratio is stated to be around 0.5, I expected larger differences then were seen in this investigation.

Another question came to mind regarding the use of powdered-iron cores for more compact inductors. Experiments with transmission line transformers showed that the red mixture, which has a permeability of 10, yields the same very high efficiency as a ferrite with a permeability of 40 (which yields the highest efficiency of practically all ferrites). **Figure 4-17** shows the results of comparisons between a coil using the red mixture and "air-wound" coils using tight and spaced windings. They all have about the same inductance. Even though the loss is excessive beyond 3 MHz, this experiment justifies its use on the 160-meter band.

An example of a very high-Q coil using 1/8-inch copper tubing is shown in **Figures 4-18** and **4-19**. As **Figure 4-19** shows, the calculated equivalent Q exceeds 900 beyond 10 MHz. Note that the Hewlett-Packard Q meter becomes unreliable above 5 MHz where the Q exceeds 600. Also shown is some very early

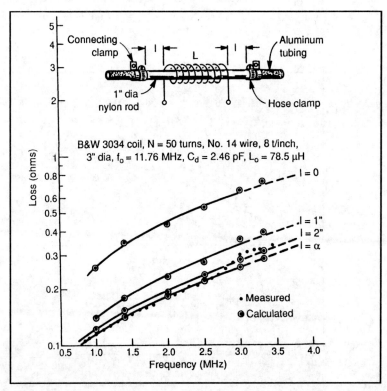

Figure 4-13. *The variation in loss with frequency for different spacings of mounting hardware (the concentric case).*

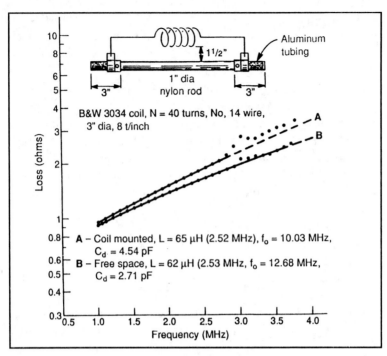

Figure 4-14 *A comparison in the loss versus frequency between a coil mounted 1.5 inches from the insulator and aluminum tubing and one in free space.*

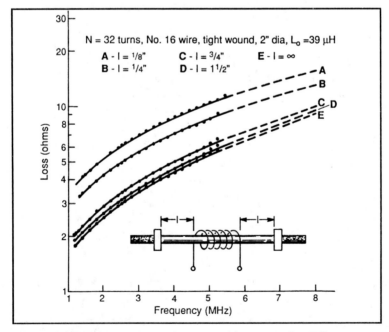

Figure 4-15. *The variation in loss versus frequency for different spacings of aluminum end-caps.*

50

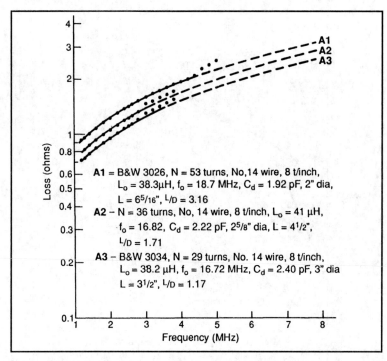

Figure 4-16. *A comparison of the loss as a function of frequency and the length-to-diameter ratio, l/D.*

data from E.L. Hall on a similar coil which was published in Terman's popular handbook.[30]

A comparison of losses was also made on rods of different insulating materials. The rods used had 1-inch diameters and sufficient lengths so the coils could be mounted concentrically without adding any loss due to the connecting hardware. The coil was a B&W 3033 of 15 spaced turns of No. 12 wire. A nylon rod and an acrylic rod showed no extra loss above the loss

Antenna	R_{in} (over low-loss ground)	R_{in} (on car)	R_{rad}	eff.(%)
RM-80	21.5	23	0.8	3.5
RM-80S	35	38	0.8	2.2
B&W 3034	8	12	0.8	6.7
RM-40	21.5	26.5	3	11.3
RM-40S	20.5	25.5	3	11.8
B&W 3034	8	14	3	21.5
RM-20	18	31	11	35.5
RM-20S	19.5	32	11	34.4
B&W 3063	15	27	11	41.0

Table 4-1. *Efficiency comparisons of mobile whip antennas.*

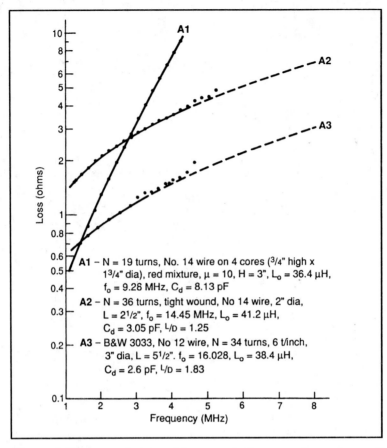

Figure 4-17. *A comparison of loss versus frequency between a coil using a powdered-iron core and two that are air-wound.*

that the coil had when it did not contain a rod. One-eighth-wall PVC and hollow phenol fiber exhibited approximately the same additional loss. The increase was about 1 percent at 5 MHz and about 2 percent at 10 MHz. Surprisingly, a solid phenol-fiber rod had a 10-percent increase in loss at 5 MHz and a 23-percent increase at 10 MHz. A dry wooden dowel and a 1/8-wall cardboard tube had approximately the same increases in loss: 4.5 percent at 5 MHz and 10 percent at 10 MHz. However, the wooden dowel, when wet, was something else. The increase in loss at 5 MHz was 92 percent, and the increase at 10 MHz was 340 percent!

Closing Comments

Just as my curiosity concerning the losses in Hustler whip antennas started me on a rather long investigation of the ubiquitous

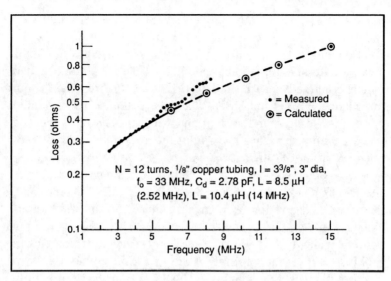

Figure 4-18. *The variation in loss with frequency for a low-loss coil using 1/8-inch copper tubing.*

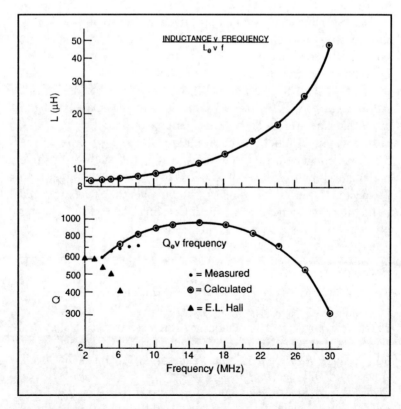

Figure 4-19. *The variation in equivalent inductance and Q with frequency of the low-loss coil of Figure 4-18 using copper tubing.*

loading coil, it is only appropriate to end this chapter with a perspective on mobile whip antennas. **Table 4-1** shows how the Hustler antennas compare in efficiency with whip antennas using low-loss loading coils, mounted on the back end of a car, which is parked on a macadam driveway. The resonant input impedance, R_{in}, is measured when the antennas are mounted over a low-loss ground system and when they are mounted on the back end of a car. The radiation resistances, R_{rad}, were obtained from *The ARRL Antenna Book*[31] for center-loaded, 8-foot whip antennas. Although the 20-meter antennas were a little less than 7 feet in height, they were loaded at the 2/3 point and, therefore, had a radiation resistance close to that of a center-loaded, 8-footer. On 20 meters, even the hustler whips are only poorer by about 5 dB from the ideal 1/4-wave condition. When looking at the other numbers, the question comes up as to whether it's worth replacing the Hustler resonators with high-Q coils. There would only be a 3 dB improvement on 40 meters and 80 meters over the low-power unit. Even the lossy RM-80-S high-power unit is only down 5 dB from an efficient radiator of the same dimensions. Another surprise was the small increase in loss to ground in going from a low-loss ground system to the ground offered by a car.

Several other results stood out in this investigation. One was the sensitivity of the loss in a coil to the spacing between turns. Another was the sensitivity of the loss to the spacing of the mounting hardware. Because the length-to-diameter ratio, l/D, has appeared so much in the literature, it was a surprise to see the experimental results that showed it is not such a sensitive parameter. Finally, the Qs of low-loss coils are much higher than I expected. I don't think values of 800 to over 900 have appeared in the literature before. It will be interesting to see if this chapter prompts some responses from the readers.

References

28. Wheeler, "Formulas for the Skin Effect," *Proc. of the IRE*, September 1942, pages 412–424.
29. Butterworth, "The High Frequency Resistance of Toroidal Coils," *Wireless and Wireless Engineer*, Vol. 6, January 1929, page 13.
30. Terman, *Radio Engineers' Handbook*, McGraw-Hill Book Company, 1943, page 75.
31. Hall, *The ARRL Antenna Book*, 16th Edition, The American Radio Relay League, Newington, Connecticut, page 16–4.

Index

Amidon Associates, Inc...38
Antenna efficiency...26–29
    and whip antennas...37
    of shortened verticals with abbreviated radials on the Earth's surface...29–32
    on-the-air comparisons...32–33

Baluns
    choke...38
    current...38
    voltage...38
Beach umbrella antenna...19–21
Brown, Lewis, and Epstein...3–4, 8
Butterworth...47, 54

Capture cross-section...11
Core, powdered-iron...49

Dipole antennas...10

End-cap
    effect...46–47
    metal...46, 48
        aluminum...48
        copper...48
        steel...48
    no end-caps...48

Fractional ratio Ununs...7, 33

Hall, E.L....51
Helical antenna...13, 16–17
Hustler resonators...37–38
Hustler whip antenna...37, 43, 52, 54

Input resistance...11–12

King...10, 22

L-C matching network...7
    pi...16–17
Loading, antennas
    base...13–14
    distributed...13, 15
    inductive...13
    midpoint...13–14, 30
    three-quarter point...13–14
    top-hat...13–17, 29–30
Loading coils...11, 15–16, 21
    air wound...49, 52
    B&W...38, 44–48, 51
    base...11
    high-Q...38, 46, 49
    homemade...38
    loss...15
        end-cap effect...46–47
        function of length-to-diameter ratio...48–49, 51
    proximity effect...42
    Q values and measurement...37–42, 44, 46–49
    skin effect...41
    theoretical considerations...38–43

wire...46–47
40-meter coils for mobile applications...47

Pi network...16–17
Proximity effect...42

Radial systems
    abbreviated radial systems...28–32
    buried...2–4, 24
    efficiency...26–29
    ground-radial system for verticals...1–8
    short ground-radial systems for short verticals...23–35
    soil conductivity and...24–26, 33

Radiation resistance...12–13, 15

Soil conductivity and ground-mounted verticals...24–26, 33
Square-root dependency...42

Top hats...13–18, 21, 32–33
    and abbreviated radial systems...29–30
    loading...13–18, 21, 29–30, 31–33
Toroids, powdered-iron...33, 49, 52
Transmission line transformers...49
    book by W2FMI...37

Ununs
    fractional ratio Unun...7, 33
    2:1 Unun...32
    4:1 Unun...7, 16–17, 33
    16:1 Unun...17, 20

Vertical antennas
    center-fed verticals...21
    ground-image vertical...2, 9
    ground-plane vertical...2
    ground-mounted vertical...1, 5, 10–11, 15, 21, 23–24, 33
    ground-radial system for...1–8
    ground radials versus elevated radials...2–3
    multiband verticals...34
    parasitic vertical beam...9, 16
    radial systems...1–8
    short ground-mounted verticals...9–22
        40 and 80-meter short vertical designs...15–21
        short ground-radial systems for...23–35
            abbreviated radial systems...29–32
            antenna efficiency considerations...26–29
            soil conductivity...24–26
    short vertical antennas for mobile use...37–54
    triband trap-vertical...7
    W2FMI Ground-Image Vertical...9
    W2FMI Ground-Mounted Short Vertical...10
    W2FMI 20-Meter Vertical Beam...10
    1/4-wavelength vertical...3–7, 17–18, 21, 23–24, 27–33
    1/8-wavelength vertical...17–18, 32
    5/8-wavelength vertical...6–7
    1/16-wavelength vertical...17, 32–33

Waltz, M.C....26
Whip antennas, Hustler...37, 43, 52, 54
W2FMI Ground-Image Vertical...9
W2FMI Ground-Mounted Short Vertical...10
W2FMI 20-Meter Vertical Beam...10

Yagi antennas...1, 10